Gifted & Talented®

To develop your child's gifts and talents

Science Experiments

For Ages 6–8

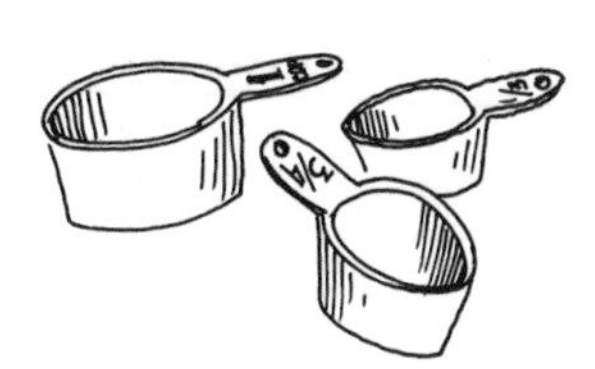

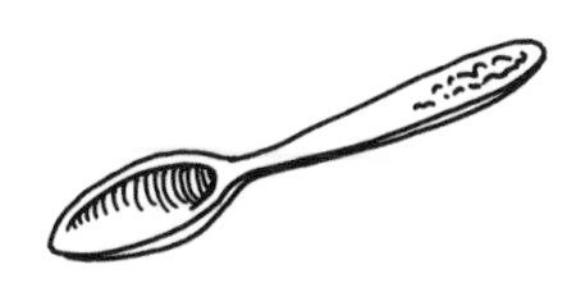

By Barbara Saffer, Ph.D.

Interior Illustrations by Leo Abbett

Cover Illustration by Larry Nolte

LOWELL HOUSE JUVENILE

LOS ANGELES

NTC/Contemporary Publishing Group

To my little scientists, Brandon and Chelsea
—B.S.

Published by Lowell House
A division of NTC/Contemporary Publishing Group, Inc.
4255 West Touhy Avenue, Lincolnwood (Chicago), Illinois 60646-1975 U.S.A.

Managing Director and Publisher: Jack Artenstein
Director of Publishing Services: Rena Copperman
Editorial Director: Brenda Pope-Ostrow
Editor: Joanna Siebert
Designer: Victor Perry

Lowell House books can be purchased at special discounts when ordered in bulk for premiums and special sales. Contact Customer Service at the above address, or call 1-800-323-4900.

Printed and bound in the United States of America

Library of Congress Catalog Card Number: 98-75617

ISBN: 0-7373-0139-2

10 9 8 7 6 5 4 3 2 1

Contents

Note to Parents 4
Introduction 5
Fast Fall 8
Nimble Nickels 10
Coin in a Cup 12
Spin the Egg 14
Spoon Sounds 16
Push the Paper 18
Puncture the Potato 20
Crazy Quarter 22
States of Matter 24
Float and Sink 26
Ocean in a Bottle 28
Peculiar Pencil 30
Color Waves 32
Sky Colors 34
Color Combos 36
Remarkable Marker 38
Temperature Trouble 40
Cool Clothes 42
Bend the Bone 44
Colorful Celery Stalks 46
Making Magnets 48
Electric Balloons 50
Water Ring 52
Scatter the Pepper 54
Fungus Fun 56
Goopy Goo 58
Answers 61
Index 64

Note to Parents

Teach a child facts and you give her knowledge. Teach her to think and you give her wisdom. This is the principle behind the entire series of Gifted & Talented® materials. And this is the reason that thinking skills are stressed widely in classrooms throughout the country.

Gifted & Talented® Science Experiments offers a variety of hands-on activities for children and adults to do together. Discuss the experiment with your child before you begin. You may need to explain some of the science concepts more thoroughly or skip over anything your child is not ready for. Your child does not need to complete all the activities that are presented with an experiment. Encourage your child to ask questions about the material.

Follow-up questions accompany each experiment to promote the development of critical and creative thinking skills. These skills include knowledge and recall, comprehension, deduction, inference, sequencing, prediction, classification, analyzing, problem solving, and creative expansion. Give your child time to think! Pause at least 10 seconds before you offer any help in answering questions. You'd be surprised how little time many parents and teachers give a child to think before jumping right in and answering a question themselves. Suggested answers, to help guide your child, are provided at the back of the book.

Gifted & Talented® Science Experiments has been written and endorsed by educators. This title will benefit any child who demonstrates curiosity, imagination, a sense of fun and wonder about the world, and a desire to learn.

Introduction

You may not realize it, but science is all around you. If you put water in the freezer, it turns to ice. When you wash dirty clothes with laundry detergent, they get clean. If you pop a balloon, it collapses. These are all examples of science at work.

Scientists want to know how and why things happen. To find the answers to their questions, scientists use a procedure called the scientific method. This consists of five steps. The following example will show you how it works.

Step 1 Ask a question.
Question: If a person throws 10 objects into the air, will some fall to Earth and some fly into space?

Step 2 Try to guess the answer. Scientists call such a guess a hypothesis.
Hypothesis: Heavy objects will fall to Earth and light objects will sail off into space.

Step 3 To check the hypothesis, set up a test or experiment.
Experiment: Throw the following 10 objects into the air: a basketball, feather, book, pencil, rock, cotton ball, bag of potatoes, leaf, baseball, and cookie.

Step 4 Observe and record the results of the experiment.
Observations: All 10 objects fall back to Earth.

Step 5 Draw a conclusion. What does the experiment show?
Conclusion: Both heavy and light objects fall back to Earth. No objects fly into space.

In this example, the results don't agree with the hypothesis. The hypothesis is wrong. The scientist would have to form a new hypothesis and do another experiment to check it.

Of course, scientists today wouldn't need to do this experiment. They already know that Sir Isaac Newton, a famous English scientist, described gravity in 1684. Gravity is a force that pulls all objects toward Earth. Gravity prevents objects from escaping into space. Objects that do fly into space, like rocket ships, must travel very fast to overcome gravity.

The experiments in this book demonstrate scientific principles (laws or facts of nature) that were discovered using the scientific method. These experiments are fun for kids and adults to do together.

In science, it's important to set up and observe experiments carefully. This often requires measuring things: size, weight, volume, time, temperature, and so on. To make accurate measurements, scientists use special instruments such as rulers, scales, measuring spoons, measuring cups, clocks, thermometers, and so forth. If you do an experiment that requires measurements, make sure that you use the proper equipment and measure as carefully as you can.

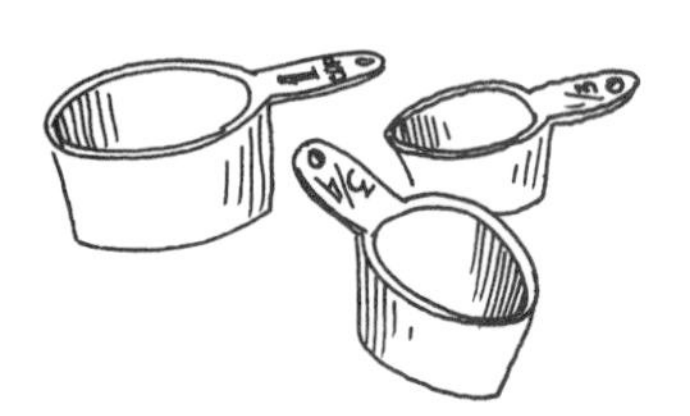

Converting to Metric

Use this chart if you need to convert a measurement to metric. For example: If an experiment calls for 2 tablespoons water and you need to measure the water using milliliters, multiply 2 by 14.8. You need 29.6 milliliters water to complete the experiment.

IF YOU HAVE	BUT YOU NEED	MULTIPLY BY
inches	centimeters	2.54
feet	meters	0.305
pounds	kilograms	0.45
ounces	milliliters	30.0
tablespoons	milliliters	14.8
teaspoons	milliliters	4.9
cups	milliliters	236.6
pints	liters	0.47
quarts	liters	0.95
gallons	liters	3.78

When you do one of the experiments in this book, be sure that you choose one that interests you. Follow these guidelines when you do an experiment:

1. Make sure that you have enough time to complete the experiment.
2. Gather all the materials you will need for the experiment before you begin.
3. Read the instructions carefully.
4. Pay attention to what is happening in the experiment.
5. Ask questions about parts of the experiment that you do not understand and discuss the results with an adult.
6. *Never* eat or drink any part of an experiment.
7. Clean up thoroughly after the experiment is done.

Sometimes an experiment may not turn out the way you expected—or the way it is described in "What Happened." If an experiment doesn't work, try it again. Now, get started and have fun!

Key Word Pronunciations

adhesion (ad-HEE-zhun)

capillary (KA-pih-LAIR-ee)

cohesion (koh-HEE-zhun)

colloids (KAH-loyds)

fungi (FUN-ji)

hypothesis (hi-POTH-eh-sis)

inertia (in-ER-sha)

ions (EYE-ons)

molecules (MOLL-eh-kyools)

momentum (moh-MEN-tum)

phloem (FLO-em)

refraction (ree-FRAK-shun)

repel (ree-PEL)

vacuum (VA-kyoom)

xylem (ZI-lum)

Nimble Nickels

Can momentum be transferred, or moved, to a coin that is not moving?

Background:

When an object is moving, it has a type of energy called momentum. The object's momentum depends on its speed and its mass. A large, fast-moving marble has more momentum than a small, slow-moving marble. Momentum can be transferred, or moved, from one object to another. If a speedy marble hits a slow marble, energy will be transferred from the speedy marble to the slow marble. The speedy marble will slow down, and the slow marble will speed up.

What You'll Need:

- eight nickels
- flat surface

MASS

All substances are made of particles called molecules. Things with more molecules, or larger molecules, have greater mass than things with fewer molecules, or smaller molecules. Therefore, a truck has more mass than a motorcycle and a rhinoceros has more mass than a chipmunk.

What to Do:

1. Place seven nickels in a row on the flat surface. Make sure the nickels touch each other.
2. Place another nickel a short distance apart from the row.

3. Hold your thumb over the tip of your index finger in front of the single nickel. Flick the nickel with your index finger so that it hits the bottom of the row. What happens?
4. Repeat the experiment with six coins in a row, and two coins apart from the others. Flick the two coins at the row. What happens?
5. Do the experiment again, with five coins in a row, and three nickels apart from the others. What happens?

6. Now repeat steps 1 through 5, flicking the coins harder this time. What happens?

What Happened:

When you flicked one nickel at the row of coins, the momentum of the moving nickel was transferred, or moved, to the row. The momentum traveled through the row, to the top nickel. The top nickel slid off. When you flicked two nickels at the row of coins, the momentum of two moving nickels traveled through the row, so the top two nickels slid off. The transfer of momentum worked the same way with three nickels. In the next part of the experiment, you flicked the coins harder. When you flicked one nickel harder, the nickel at the top of the row slid off faster and farther than before. Each time you did this, the coins moved faster and farther, but the number of coins that moved was always the same as the number of coins that you flicked at the row. In each part of this experiment, the momentum of the coins that slid off at the top was exactly the same as the momentum of the coins that hit the row.

One Step Further:

Line up five pennies in a row. Make sure the pennies are touching each other. Place one quarter a short distance apart from the row. Flick the quarter with your finger so that it hits the bottom of the row. Two pennies slide off the top of the row. Why? Try the experiment with different combinations of coins and see what happens.

Questions:

1. Does a book lying on your desk have momentum?
2. Which has greater mass, an elephant or a mouse?
3. If a car speeds up, does its momentum increase?
4. What do you think has greater momentum: a big train traveling at 50 miles per hour or a small car traveling at 50 miles per hour? Why?
5. Can you describe a situation in which a person transfers momentum to another person?

Coin in a Cup

How can you move a coin without touching it?

Background:

If you put a coin on a table, it will stay there until someone or something moves it. We say the coin has inertia. Every object in the universe has inertia. This means that an object that is not moving will stay that way until something moves it. An object that is moving will keep moving in the same direction until something stops it.

What You'll Need:

- small plastic cup
- flat surface
- 3-by-5-inch index card
- coin (nickel or quarter)

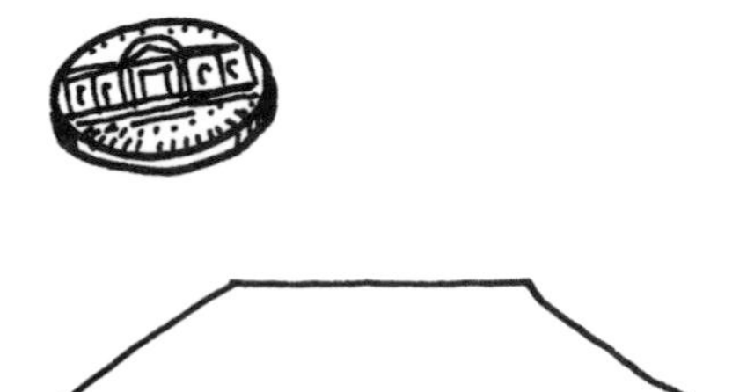

What to Do:

1. Place the cup on a flat surface.
2. Center the index card over the top of the cup.
3. Place the coin in the middle of the card.
4. Quickly flick your index finger against the edge of the card. What happens?

What Happened:

At the beginning of the experiment, the cup, the card, and the coin were all motionless. Because of inertia, they remained where they were. When you flicked the card with your finger, the card flew off and the coin dropped into the cup. The flick of your finger was a force, or type of power, that moved the card. You did not touch the coin or cup, so they did not move. Once the card was gone, however, the force of gravity pulled the coin down (see "Fast Fall").

One Step Further:

Place a plastic cup on its side on a flat surface. Put a marble into the cup. Quickly push the cup forward about a foot, then stop it suddenly. The marble will roll out of the cup. Why does that happen?

Questions:

1. What do you think would have happened if the coin had been glued to the card?
2. When you're riding in a car that stops suddenly, your backpack, books, or other items may slide off the seat. Why does this happen?
3. Because of inertia, an object that is moving keeps moving until something stops it. Why do you think the index card stopped moving shortly after you flicked it off the cup?
4. Can you name a moving object in the universe that will keep moving for a very long time?
5. What do you think would happen if objects in the universe lost their inertia?

Spin the Egg

Can you tell the difference between a raw egg and a hard-boiled egg by spinning them?

Background:

If an object is spinning, it will keep spinning until someone or something stops it. This is because of the object's inertia (see "Coin in a Cup"). For example, if you start a top spinning, then press your finger on it, the top will stop. Your finger acts as a force, or power, that stops the spinning top.

What You'll Need:

- raw egg
- flat surface
- cooled hard-boiled egg

What to Do:

1. Start the raw egg spinning on a flat surface. Let it spin for a few seconds.
2. Stop the spinning egg by touching it lightly with your finger. As soon as the egg stops, remove your finger. What happens?
3. Now spin the hard-boiled egg on a flat surface. Let it spin for a few seconds.
4. Stop the spinning egg by touching it lightly with your finger. As soon as the egg stops, remove your finger. What happens this time?

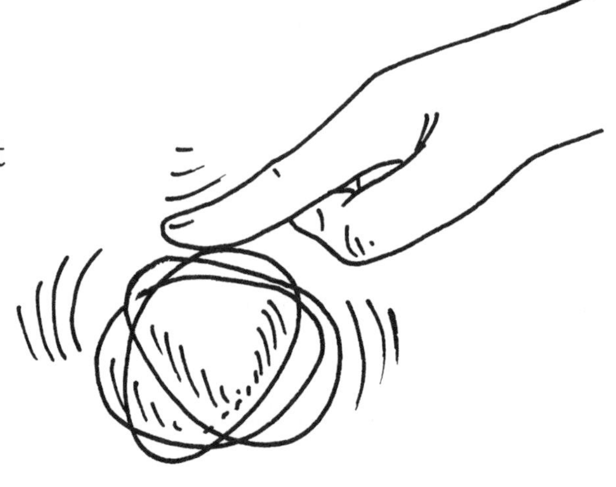

What Happened:

In a raw egg, the yolk is liquid, so it is not connected to the shell. When you touched the raw egg, you made the shell stop spinning. However, the yolk inside kept twirling because of inertia. When you removed your finger from the egg, the movement of the twirling yolk made the egg start spinning again. A hard-boiled egg is completely solid. When you touched the hard-boiled egg, the entire egg—including the yolk—stopped spinning. The egg remained still when you removed your finger.

One Step Further:

Place a hard-boiled egg and a ball on the floor. Roll each one. What happens?

Questions:

1. What will happen if you spin a solid rubber ball, then stop it? Will the ball start spinning again? Why or why not?
2. The needle of a compass spins around to point in a certain direction. Can you name the direction?
3. Even if you don't touch a spinning top, it stops after a while. Why is that?
4. Can you name an object in our solar system that spins?
5. Many tasty things, including omelets, cake, and cookies, are made with eggs. On a separate piece of paper, draw pictures of two foods you like that contain eggs. Can you name some other items that are used to make these foods?

Spoon Sounds

What happens when sound waves travel through a string?

Background:

When something produces a sound, it makes air molecules, or particles, vibrate, or move, forward and back. The air molecules hit other air molecules, making them vibrate. This goes on and on, so that the sound is carried along as "sound waves." Sound waves don't travel only through air, which is a gas. They can also move through solids and liquids by making the molecules vibrate back and forth.

What You'll Need:

- metal spoon
- table
- 2 feet of thin string

What to Do:

1. Pick up the spoon and tap it on the table. What do you hear?
2. Now place the spoon in the center of the string, and tie the string around the handle of the spoon.
3. Wrap one end of the string around your right index finger. Wrap the other end of the string around your left index finger.
4. Carefully place the tips of your index fingers in your ears.
5. Lean forward so that the spoon taps against the table. What do you hear now?

What Happened:

Sound waves are carried along as molecules bump into other molecules. Substances with molecules that are close together carry sound waves better than substances with molecules that are far apart. In solid objects, the molecules are close together. Gases, such as air, have molecules that are farther apart. When you held the spoon and tapped it on the table, the sound waves moved through the air to your ears. You heard muffled "clinks." When you attached the spoon to a string and tapped it against the table, the sound waves traveled up the string to your ears. Since the string is a solid, and therefore carries sound waves better than air does, you heard loud "dings."

One Step Further:

To make a paper tube kazoo: Get a paper tube from an empty roll of paper towels or toilet paper. Fasten a small piece of waxed paper over one end of the tube with a rubber band. Place the open end of the tube near your mouth and hum loudly into the tube. Keep your lips parted while you hum. The tube will make a funny buzzing sound. Why is that?

Questions:

1. Do most of the sounds you hear come through solid objects or through air?
2. What are some sounds you might hear in your home? What are some sounds you might hear in the street?
3. Do you think sound waves are stronger in water or in air?
4. Sound waves cannot travel through outer space, because there is no atmosphere, or layer of air, as there is on Earth. Two astronauts fixing the outside of their spaceship wouldn't be able to speak to each other. How could they communicate?
5. People in different parts of the world speak different languages. However, some words are similar in many languages. For example, the fruit called a pear (PARE) in English is *pera* (PERR-ah) in Spanish, and *poire* (PWAR) in French. Why do you think different languages have some words that sound alike?

Push the Paper

Can air pressure hold down a sheet of newspaper?

Background:

A layer of air, hundreds of miles thick, covers the Earth. This is called the atmosphere. The air is made of molecules, or small particles, which bounce around—up, down, and sideways—pushing on everything they touch. The pushing force of the air molecules is called air pressure. Air pressure is greatest near the Earth's surface because the air molecules are squeezed together by many miles of atmosphere above. Farther up, the air molecules are not squeezed together as much, and air pressure is not as strong.

What You'll Need:

- large sheet of newspaper
- table
- 12-inch ruler

What to Do:

1. Spread the sheet of newspaper so that the bottom edge of the paper runs along the edge of the table. Smooth the paper so it lies flat.
2. Slide the ruler under the center of the newspaper so that 3 inches of ruler sticks out from the edge of the table.
3. With your hand, quickly slap the uncovered end of the ruler. Does the newspaper lift up?
4. Arrange the newspaper and ruler again. This time, slowly press your hand down on the uncovered end of the ruler. Does the newspaper lift up this time?

What Happened:

You can't lift the newspaper by slapping the ruler because air pressure is pushing down hard on the paper. Even if you are very strong, it's difficult to lift the newspaper against air pressure this way. When you press on the ruler slowly, however, air gets under the paper. The air beneath the paper pushes up, while the air above the paper pushes down. Because the pressure is now the same on both sides of the paper, the ruler can lift the paper easily.

One Step Further:

Put a straw into a glass of water. Suck some water into the straw, then quickly cover the top of the straw with your finger. Lift the straw out of the water in the glass. The water in the straw remains there. Now lift your finger off the top of the straw. The water in the straw flows out. Why does that happen?

Questions:

1. Why can you "see" air more easily on a windy day?
2. Air pressure pushes on the front of your body. Why doesn't it stop you from walking forward?
3. How do you think suction cups work?
4. If a layer of water, hundreds of miles thick, covered the Earth instead of air, would there be more or less pressure at the Earth's surface? Why?
5. A barometer is a tool used to measure air pressure. How do you think a pilot could use a barometer to measure altitude, or how high the plane is flying?

Puncture the Potato

Can a straw containing compressed air make a hole in a potato?

Background:

Compressed air is air that is squeezed into a small space. In compressed air, the air molecules, or particles, are crowded together. Compressed air, therefore, has higher pressure, or pushing power, than normal air.

What You'll Need:

- raw potato
- flat surface
- two plastic drinking straws

What to Do:

1. Place the potato on a flat surface.
2. Hold one of the straws straight up, about 6 inches above the potato. Don't cover either opening of the straw.
3. Hold the potato tightly and try to stab it with the straw. Does the straw puncture the potato? Why or why not?
4. Now hold the second straw straight up, about 6 inches above the potato. This time, tightly cover the top opening of the straw with your thumb.
5. Try to stab the potato again. Does the straw puncture the potato this time? Why or why not?

What Happened:

The first straw couldn't puncture the potato because it didn't contain compressed air. When you covered the second straw with your finger and stabbed the potato, the air in the straw could not escape. Instead, the air was compressed, or pushed into a small space, as the straw moved downward. Compressed air has higher pressure, so it pushed against the inside of the straw. This made the straw stiff enough to punch a hole in the potato.

One Step Further:

Air can be used to make music. Pour water into an empty soda bottle until the bottle is about three-quarters full. Hold a straw inside the bottle. Blow across the top of the straw. You will hear a sound. Now continue to blow as you move the straw up and down in the bottle. You will hear different sounds. Why?

Questions:

1. What does it mean when something is compressed?
2. Can you name an object in your home that can be compressed?
3. If you had used a baked potato instead of a raw potato, would it have been easier or harder to poke a hole in it? Why?
4. It's easier to compress a gas than a liquid or a solid. Why do you think that is?
5. Compressed air is used to run compressed-air motors like those in jackhammers. Can you name a machine that uses a gasoline-powered motor? Can you name a machine that is powered by electricity? Imagine that you invented a "homework-helper machine." What would you use to power it?

Crazy Quarter

Can air make a quarter dance?

Background:

When air gets cold, it contracts, or takes up less space. When air gets warm, it expands, or takes up more space.

What You'll Need:

- empty 2-liter soda bottle without a cap
- freezer
- flat surface
- water
- quarter

What to Do:

1. Put the empty soda bottle in the freezer for at least 30 minutes.
2. Take the bottle out of the freezer and place it on a flat surface.
3. Wet the quarter and place it over the top of the open bottle. The quarter must completely cover the opening of the bottle.
4. Put your hands around the sides of the bottle to warm it. Wait a few moments. What happens?

What Happened:

When you put the bottle in the freezer, the air in the bottle became cold. The cold air contracted, or shrank. When you removed the bottle from the freezer and held it, the air inside the bottle became warmer. The warm air expanded, or spread out, increasing the pressure in the bottle (see "Push the Paper"). The expanding air pushed out and up. The quarter covered the top of the bottle, so the air pushed up on the quarter. The quarter jumped, and a little air escaped. This released some of the pressure. As the air in the bottle continued to warm up, it pushed on the quarter again and again, making the quarter "dance."

One Step Further:

Prepare a large bowl of cold water filled with ice cubes. Ask an adult to help you fill an empty 2-liter soda bottle with very hot tap water. Leave the water in the bottle for three minutes, then pour it out. Stretch the open end of a balloon over the mouth of the bottle. Put the bottle into the icy cold water. The balloon will enter the bottle and inflate. Why do you think this happens?

Questions:

1. If you blew up a balloon, tied the end, and put it in the freezer, what do you think would happen? Try it and see.
2. Why do you think the quarter had to completely cover the opening of the bottle for the experiment to work?
3. Can you name three things that expand and contract?
4. A jar with a screw-top lid is sometimes hard to open. If the jar is placed under hot running water for a minute, it may open more easily. How do you think this works?
5. If you found a magic potion that could make you grow larger or smaller, what kinds of things would you want to do?

States of Matter

Can a solid become a gas?

Background:

Everything in the universe, from the smallest ant to the largest star, is made of matter. Matter comes in three states, or forms: solid, liquid, and gas.

What You'll Need:

- three ice cubes
- zip-top sandwich bag
- microwave-safe dish
- microwave oven

What to Do:

1. Put the ice cubes into the sandwich bag. Seal the bag.
2. Place the bag on a microwave-safe dish. Put the dish inside the microwave oven.
3. Ask an adult to help you heat the bag for 90 seconds at high power. If the ice isn't completely melted, heat the bag for another 30 seconds. The ice will form water. 
4. Now heat the bag for another 30 seconds. What happens? Why?
5. Let the bag cool completely before removing it from the microwave.

What Happened:

In this experiment, matter changed its state, or form. Solid ice became liquid water, then a gas called water vapor. This happened because the molecules, or particles, of matter were heated up. The molecules of matter are always moving around. In a solid, like ice, the molecules are close together and move very slowly. When a solid is heated, the molecules move farther apart. The solid becomes a liquid. When a liquid is heated, the molecules move even faster and farther apart, becoming a gas. Your bag puffed up because the water became water vapor. The gas molecules moved farther apart, filling more space in the bag. Matter can also change its state in the other direction. When a gas is cooled enough, it can become a liquid, then a solid.

One Step Further:

When you finish the experiment, put the bag in the freezer. Check it after three or four hours. What happens to the bag? Why?

Questions:

1. Is your body made of matter?
2. If you leave a chocolate bar out in the hot sun, it melts. Why does this happen?
3. Can you name a solid? A liquid? A gas?
4. What's the name of the red-hot liquid that pours out of a volcano? What does this liquid form when it cools?
5. In science fiction stories, spaceships sometimes use "matter" and "antimatter" as fuel for their engines. When matter and antimatter collide, they release tremendous energy. Make up a science fiction story that contains the words *matter* and *antimatter*.

Float and Sink

Why does a boat float?

Background:

Some objects float on water and other objects sink in water. An object's density is how much it weighs compared to its volume, or how much space it takes up. Objects that are less dense than water (weigh less than water in the same amount of space) float. Objects that are more dense than water (weigh more than water in the same amount of space) sink.

What You'll Need:

- clay
- large bowl
- water
- flat surface

What to Do:

1. Divide your clay into two equal parts.
2. Roll one half of the clay into a ball.
3. Shape the other half of the clay into a flat-bottomed boat.
4. Put water into the bowl and place it on a flat surface.
5. Place the clay ball into the water. Does it sink or float? Why?
6. Now take out the clay ball, and put the clay boat into the water. Does it sink or float? Why?

What Happened:

In this experiment, the clay ball and the clay boat weigh about the same. The clay ball, however, has less volume, or takes up less space, than the clay boat. In other words, the ball is more dense than the boat. The ball sinks because it is more dense than water. The boat floats because it is less dense than water.

One Step Further:

Gather several different objects, such as a rock, apple, piece of paper, small plastic bowl, spoon, penny, leaf, and so on. Guess which objects will float and which will sink. Then place each object in a bowl of water and see if you guessed correctly.

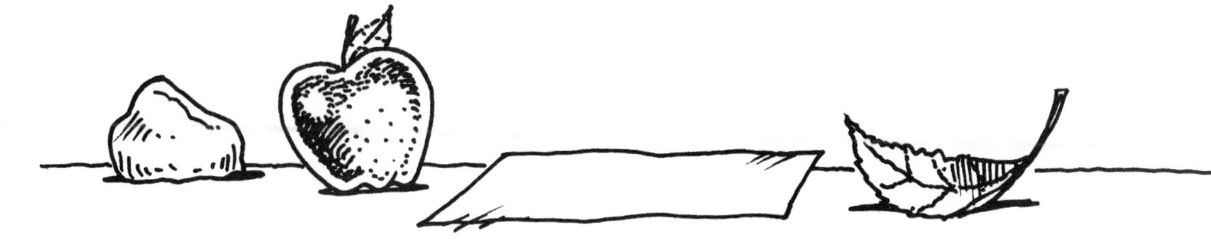

Questions:

1. Why can most people float in water?
2. Large ships can float in water while small marbles sink. Why do you think this is?
3. Water from the Dead Sea, a salt lake between Israel and Jordan, contains about seven times as much salt as water from the ocean. Do you think the Dead Sea is more or less dense than the ocean? Why?
4. The great ship *Titanic* sank after an iceberg tore a hole in its side. Why do you think the ship sank?
5. When an astronaut takes a space walk, he or she floats in space. How is this different from floating in water?

Ocean in a Bottle

Is oil more or less dense than water?

Background:

Liquids that are less dense than water float. Density is a measure of how much an object weighs compared to its volume, or how much space it takes up (see "Float and Sink"). Liquids that float weigh less than the same volume, or amount, of water.

What You'll Need:

- water
- empty, clear 2-liter plastic soda bottle with cap (label removed)
- blue food coloring
- large bottle of baby oil

What to Do:

1. Pour water into the plastic bottle until it is about three-quarters full.
2. Add 10 drops blue food coloring to the bottle, and tighten the cap. Shake the bottle to mix the food coloring and water.
3. Remove the cap from the bottle. Fill the bottle to the rim with baby oil, then replace the cap.
4. Hold the bottle on its side. The clear oil will float on the blue water.
5. While holding the bottle on its side, gently rock the bottle by tilting up one side, then the other side. The blue water will look like the waves of the ocean.

6. Now shake the bottle vigorously. The oil and water will appear to blend together. Stop shaking. Do the oil and water separate again?

What Happened:

When you poured oil and water into the soda bottle, they formed two layers. This happened because oil and water don't mix. Oil molecules, or particles, are very different than water molecules. They can't blend together. Since oil is less dense than water, the oil floated on top. When you rocked the bottle gently, the water sloshed back and forth under the oil, making the "ocean waves." When you shook the bottle, the oil and water got jumbled together, but they didn't really mix. They separated again when you stopped shaking.

One Step Further:

Get a small jar with a lid. Pour ¼ cup cooking oil, ¼ cup water, and ¼ cup honey into the jar. Tighten the lid. The liquids will form three layers. Which layer do you think will be on the bottom? Which one will be on top? Why?

Questions:

1. Why do you think the food coloring was able to mix with the water in this experiment?
2. After an accident, an oil tanker may spill millions of gallons of oil into the ocean. Where do you think the oil goes?
3. There are many kinds of oil. Can you name three things oil is used for?
4. We sometimes say two people are "like oil and water." What do you think this expression means?
5. Your hands sometimes get greasy, or oily, when you eat french fries or fried chicken. Do you think you could clean your hands just by running plain water over them? How do you think soap helps you clean your hands?

Peculiar Pencil

Can refraction make a pencil look bent?

Background:

Light can travel at about 186,000 miles per second in outer space. When light travels through a substance made of particles, however, it collides with the particles and slows down. The more dense, or heavy, the substance, the more it slows down the light (see "Float and Sink"). Air slows down light a tiny bit. Glass, which is denser than air, slows down light a bit more. Water, which is denser than glass, slows down light even further. So, when light passes from air to glass to water, it slows down quite a bit. When light goes in the other direction, from water to glass to air, it speeds up. Whenever light's speed changes, the light bends. This is called refraction. Refraction can change an object's appearance.

What You'll Need:

- water
- tall clear glass
- flat surface
- long pencil

What to Do:

1. Pour water into the glass until it's almost full.
2. Place the glass on a flat surface.
3. Put the pencil into the glass. Part of the pencil should stick out above the water.
4. Place your eyes level with the glass and look at the pencil from the side. Does it look bent?

What Happened:

For you to see an object, light must bounce off the object and enter your eyes. Light that bounced off the top part of the pencil traveled straight through the air to your eyes. Its speed stayed the same, so it was not bent. Light from the submerged, or underwater, part of the pencil traveled from water to glass to air before it reached your eyes. Because the light increased its speed when it moved from water to glass to air, it was bent. This made the submerged part of the pencil look bent.

One Step Further:

Hold a metal spoon straight up and look at your image in the bowl part of the spoon. What do you see? Now turn the spoon over and look into the back part that curves outward. How has your image changed? Why did this happen?

Questions:

1. Light bends when it passes from water to air. Does light bend when it passes from air to water?
2. Can you name five things, other than light, that bend?
3. If you used a long straw instead of a pencil, would it look bent? Try it and see.
4. Why can't you see in the dark?
5. Light travels at about 186,000 miles per second in outer space. How would you figure out how far light travels in a minute? In an hour?

Color Waves

How do the colors in sunlight form a rainbow?

Background:

Sunlight is called white light. However, sunlight is really made of all the colors of the rainbow. These colors are red, orange, yellow, green, blue, and violet. Light moves in waves, like the waves of the ocean. Each color of light has a different size wave.

What You'll Need:

- ruler
- scissors
- 3-by-5-inch index card
- tape
- tall clear glass
- water
- flat surface
- window
- sheet of white paper

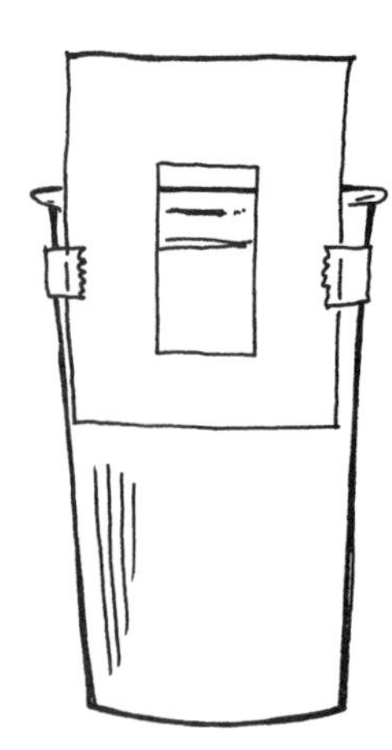

What to Do:

1. Ask an adult to help you measure and cut out a ½-by-3-inch rectangle from the center of the index card.
2. Tape the card to the glass so that the long side of the rectangle is straight up. The opening in the card should extend a little bit above the rim of the glass.
3. Fill the glass to the rim with water.
4. Set the glass on a flat surface in front of a sunny window. Sunlight should shine into the glass through the opening in the card.
5. Place the sheet of white paper in front of the glass. A rainbow will appear on the paper as the sunlight shines through the water.

What Happened:

White light is made of many colors. Each color has a different size wave. A rainbow appears when sunlight is bent, or refracted, by water drops in the atmosphere (see "Peculiar Pencil"). When sunlight is refracted, each size wave is bent by a different amount. Violet light, which has the shortest waves, is bent the most. From violet to blue to green to yellow to orange to red, the length of each wave increases, and the amount that each wave bends decreases. Red light, which has the longest waves, is bent the least. Since the colors of light are bent by different amounts, they spread out. In this experiment, the water in the glass refracted sunlight, so the colors spread out to form a rainbow.

One Step Further:

To make a color spinner: Draw a circle that measures 4 inches across on a sheet of white paper. Use the lid of a jar as a guide if you need to. Cut the circle out. Mark off six equal sections, shaped like pizza slices, on the circle. Fill in the sections with the colors of the rainbow, in order: violet, blue, green, yellow, orange, and red. Make a little hole in the center of the circle. Slip the circle over the handle of a small plastic top. Tape the bottom of the circle to the top. Spin the top. What do you see? Why?

Questions:

1. Why are you most likely to see a rainbow after it rains?
2. Which color of light has longer waves, green or orange?
3. Light is a form of energy. Can you name another form of energy?
4. Sir Isaac Newton was the first person to discover that white light is made of many colors. Can you name something else Sir Isaac Newton is famous for? Hint: Reread the introduction if you don't know.
5. Many animals are camouflaged, or hidden, because their colors blend into their habitat, or the place where they normally live. On a separate piece of paper, draw a picture of an animal that is camouflaged in its habitat.

Sky Colors

Why does the sky change color at sunrise and sunset?

Background:

Light behaves like ocean waves in many ways. When waves collide with objects in their path, they bounce off the objects and scatter in all directions. Small waves scatter more than large waves. As sunlight travels through Earth's atmosphere, the light waves collide with many obstacles, like air molecules, bits of dust, and water droplets. The light waves bounce off these particles and scatter all around. Sunlight is made of all the colors of the rainbow, but each color has a different size wave (see "Color Waves"). The shortest light waves appear blue and are scattered more than those of any other color. That's why the sky appears blue during the day.

What You'll Need:

- deep glass bowl
- water
- 2 tablespoons milk
- table
- three books
- flashlight

What to Do:

1. Fill the bowl almost to the top with water.
2. Stir in 2 tablespoons milk. The bowl of liquid represents the sky.
3. Put the bowl of liquid on a table.
4. Stack the books about 3 feet away from the bowl.
5. Place the flashlight on the books and turn it on. Make sure the flashlight is shining into the bowl. The flashlight represents the Sun.

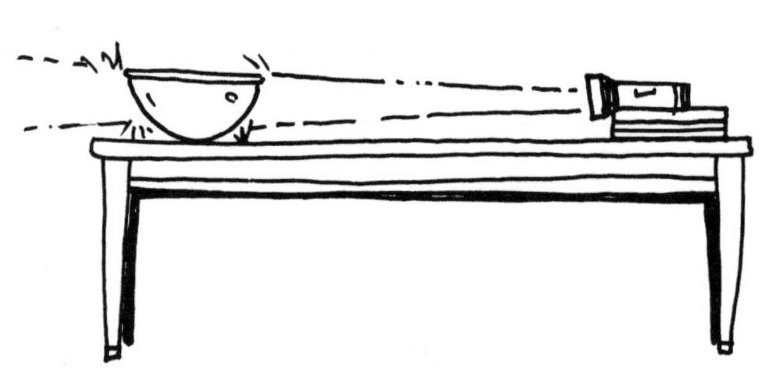

6. Darken the room.
7. Stand on the opposite side of the bowl from the flashlight. Crouch down and gaze into the bowl. What color is the "sky"?

What Happened:

When the flashlight beam passed through the liquid in the bowl, the light waves struck a huge number of particles. The short waves of blue light were scattered so much they could not pass through the bowl. The longer waves of red light were not scattered as much and were able to pass through the bowl. Therefore, the "sky" in the bowl looked red. This is similar to what happens to the real sky at sunrise and sunset. At these times of day, the Sun is low in the sky. Sunlight must pass through more atmosphere, and more particles, than at midday. The thick atmosphere "stops" the blue light but lets the red light pass through, making the sky look red.

One Step Further:

Get two sheets of white paper and a pink crayon. Lay the two sheets of paper on a table, side by side. On one sheet of paper, draw a pink circle, about 3 inches across. Color the circle so that it's solid pink. Now stare at the pink circle while you slowly count to 30. Then quickly look at the sheet of plain white paper. What do you see? Why does this happen?

Questions:

1. Why is the sky black at night?
2. An apple looks red because red light bounces off it. Grass looks green because green light bounces off it. Think of about five objects in your home or school. What color light bounces off each object?
3. How do you think water and dust get into the atmosphere?
4. Can you name another kind of wave other than light waves?
5. What color is the sky at your favorite time of day? On a separate piece of paper, draw a picture of what you like to do at your favorite time of day.

Color Combos

What colors can be made by mixing primary colors?

Background:

Paint, crayons, markers, and dyes come in many colors. All these colors are made from different combinations of just three colors—red, yellow, and blue. These three colors are called the primary colors. When two primary colors are combined, the color that results is called a secondary color.

What You'll Need:

- white construction paper
- flat surface
- three plastic spoons
- red, yellow, and blue washable paints
- medium-sized paintbrush
- small bowl of water
- paper towel

What to Do:

1. Place the construction paper on a flat surface.
2. Put a spoonful of each color paint near the top of the construction paper. Keep the colors separate.
3. Pick up a small glob of red paint with the paintbrush, and dab it on the paper a few inches below the spoonfuls of paint. Wash the paintbrush in the bowl of water and dry it on the paper towel.
4. Now pick up a small glob of yellow paint with the paintbrush and mix it with the glob of red paint. What color paint does this make?
5. Wash and dry the paintbrush again.

6. Repeat steps 3 and 4 using red and blue paint. What color paint does this make?
7. Wash and dry the paintbrush again.
8. Repeat steps 3 and 4 using blue and yellow paint. What color paint does this make?

What Happened:

When you mix two primary colors, you make a secondary color. Red and yellow make orange. Red and blue make purple. Blue and yellow make green. Orange, purple, and green are secondary colors. You can make many colors, including black, by mixing different amounts of primary and secondary colors. To make lighter colors, just add a little white paint. With red, yellow, blue, and white paint, you can make any color you need.

One Step Further:

Dip your thumb into one of the colors you made. Then press your paint-covered thumb onto a fresh sheet of white paper. This will create a colorful thumbprint. Make several thumbprints, using different colors. Let the paint dry. Then use markers to add features, like faces, arms, legs, tails, and antennae to your thumbprints. Create a page full of zany thumbprint creatures. What are the names of your creatures?

Questions:

1. What are the three primary colors? What are the three secondary colors?
2. Which colors would you mix to make lavender, or light purple, paint?
3. Can you name two fruits or vegetables that change color as they get ripe?
4. What do you think it means when someone is “color blind”?
5. Pretend you are a clothing designer. On a separate piece of paper, draw an outfit containing two colors that go well together. Why do you think certain colors go well together?

Remarkable Marker

Can the ink of a black marker be separated into different colors?

Background:

If you color a picture with a black marker, the picture looks black. However, the black ink in the marker is actually made of several colors mixed together.

What You'll Need:

- scissors
- white paper towel
- ruler
- tall clear glass
- nontoxic washable black marker
- water
- flat surface
- pencil
- tape

What to Do:

1. Cut a strip from the paper towel. Make the strip 1 inch wide and almost as tall as the glass.
2. Color a large dot with the black marker, about 2 inches from the bottom of the paper strip.
3. Pour an inch of water into the glass and put the glass on a flat surface.
4. Fold the top of the paper strip over the center of the pencil and tape it, so it can't slip off.
5. Place the pencil on top of the glass so that the paper strip hangs into the glass. Make sure the bottom inch of the paper strip is in the water.
6. Watch the water creep up the paper. What happens to the black ink?

What Happened:

When the paper strip was dipped into the water, the water crept up the paper. This process is called capillary action. As the water moved up the paper, it dissolved, or broke up, the ink from the black marker. The ink in the black marker is composed, or made, of several colors. Some colors dissolve better in water than others. The colors that dissolve well travel far up the paper. The colors that dissolve poorly travel only a short distance up the paper. After a few minutes, all the colors are separated. Black ink often separates into shades of purple, blue, and green. The exact colors, however, depend on the type of marker used.

One Step Further:

Repeat the experiment with different colored markers (red, green, blue, orange, and so on). Which markers are made with colors that split into other colors? Which are not?

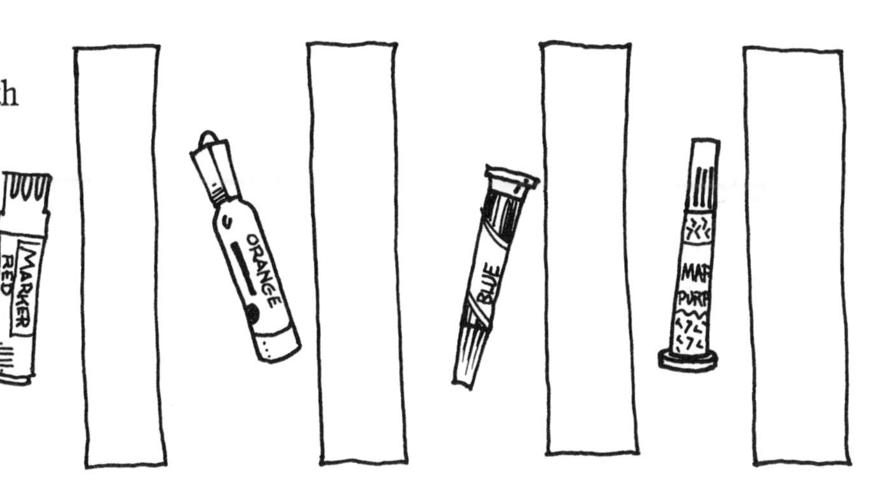

Questions:

1. What color in the black ink dissolved best in water?
2. People have used plants to color cloth for thousands of years. Which colors do you think come from beets, grapes, spinach, onion skins, and blueberries?
3. What happens when you place a sponge on a small puddle of water? What do you think this process is called?
4. Some boxes of crayons have 96 colors. How do you think these colors are produced?
5. Pretend a marker company has asked you to create five new colors that look and smell like fruits and flowers. What would your new colors look like? What would you call them?

Temperature Trouble

Can your hands always judge the temperature of water?

Background:

To find out if something is hot or cold, you can measure its temperature with a thermometer. You can also feel it with your hand. If you touch a rock that's been lying in the hot sun, it feels warm. If you pick up an ice cube, it feels cold. People often use their hands to judge temperature.

What You'll Need:

- three large bowls
- water
- ice cubes
- flat surface

What to Do:

1. Fill the first bowl with water and ice cubes, so that the water is very cold.
2. Fill the second bowl with lukewarm water.
3. Ask an adult to help you fill the third bowl with hot tap water. HAVE AN ADULT MAKE SURE IT'S NOT HOT ENOUGH TO HURT YOU.
4. Line the bowls up on a flat surface, with the hot water on the left side, the lukewarm water in the center, and the ice water on the right side.
5. Place your left hand into the hot water and your right hand into the ice water. Slowly count to 30 while you keep your hands in the water.

6. Now quickly remove your hands from the hot and cold water, and place both hands in the lukewarm water. Does one of your hands feel cool? Does one of your hands feel warm? If so, which one feels cool and which one feels warm?

What Happened:

When you put your hands in the lukewarm water, your hands did not really feel the temperature of the water. They felt the difference in temperature between the lukewarm water and the water they had been in. When your left hand was in hot water, it felt hot. The lukewarm water was cooler, so it made your left hand feel cool. When your right hand was in cold water, it felt cold. The lukewarm water was warmer, so it made your right hand feel warm. This experiment shows that your hands can't always judge the temperature of water.

One Step Further:

Prepare two bowls of icy cold water. Put a glove on one hand, and leave the other hand bare. Place a plastic bag over each hand. Ask an adult to fasten the bags at your wrists with loose rubber bands. Put each hand into a bowl of icy water. Be sure to keep the tops of the bags out of the water. Which hand feels warm longer? Why?

Questions:

1. Can you name five foods that taste best when they're hot? Can you name five foods that taste best when they're cold?
2. When you get sick, the doctor may take your temperature. What instrument does he or she use to do this? What is the doctor trying to find out?
3. Can you ever tell if something is hot or cold by looking at it? What kind of clues could you use?
4. Your sense of touch helps you tell hot from cold. Can you name your other four senses?
5. On a separate piece of paper, draw a picture of a thermometer on a hot day and a thermometer on a cold day. How are they different?

Cool Clothes

Can the color of your clothes help you feel cool?

Background:

Light may look colorless, but it is actually made of many colors mixed together (see "Color Waves"). When light hits an object, some of the colors are absorbed, or taken in. Other colors are reflected, or bounced off. The colors that are reflected give an object its hue, or particular color. For example, a banana looks yellow because it reflects yellow light. An object that reflects all the colors of light appears white. An object that absorbs all the colors of light appears black. Light is a form of energy. When objects absorb light, they heat up.

What You'll Need:

- black sock
- white sock
- bright lamp or sunlight

What to Do:

1. Put the black sock on your left hand and the white sock on your right hand.
2. Hold your hands up a few inches in front of a lamp or in direct sunlight.
3. Slowly count to 100, but stop if it becomes uncomfortable. Does one hand feel warmer than the other?

What Happened:

The black sock absorbed all the light that struck it. Since light has energy, the black sock got hot. This made your left hand feel warm. The white sock reflected all the light that struck it. The white sock didn't absorb light energy and didn't get as hot. Your right hand, therefore, didn't get as warm as your left hand. That is why white clothes will keep you cooler than black clothes on a summer day.

One Step Further:

Make a batch of Sunlight S'mores. Put four graham crackers side by side in the bottom of a glass baking pan. Place a chocolate bar on top of two of the graham crackers. Put six mini-marshmallows on top of each of the other two graham crackers. Cover the baking pan with clear plastic wrap. Put the pan in the hot sun. The plastic wrap and glass baking pan will "trap" the sunlight's heat. After the chocolate bar and marshmallows melt, put the graham crackers together to make Sunlight S'mores. Eat and enjoy!

Questions:

1. Which colors of light are reflected by a carrot, a beet, a white pearl, and a black marble?
2. How do black clothes help you keep warm on a cold winter day?
3. How do mittens—even white ones—help keep your hands warm in winter?
4. Why shouldn't you look directly at the Sun?
5. Places that have hot weather most of the time are said to have a hot climate. Can you name three places with a hot climate? Places that have cold weather most of the time are said to have a cold climate. Can you name three places with a cold climate?

Bend the Bone

Can you tie a knot in a bone?

Background:

Bones are made of two main substances: hard minerals that contain calcium, and soft proteins. If a bone's calcium is removed, the soft proteins left behind make the bone stretchy, like a very thick rubber band.

What You'll Need:

- large drumstick from a cooked chicken
- water
- scrub brush
- large glass jar with a lid
- white vinegar
- flat surface

What to Do:

1. Ask an adult to save a drumstick from a chicken. Remove the meat from the drumstick. Clean the bone thoroughly with water and a scrub brush. Let the bone dry for a few hours.

2. Fill the jar almost to the top with vinegar. Put the jar on a flat surface. Place the bone inside the jar and tighten the lid. Let the jar sit for three days.

3. After three days, remove the bone from the vinegar and rinse it with water. The bone should be soft and stretchy. Try to tie it in a knot. If the bone isn't soft enough yet, prepare a fresh jar of vinegar and put the bone back in. Check the bone every two or three days, until it is soft enough to tie in a knot.

What Happened:

Vinegar dissolves, or breaks up, the hard calcium minerals in the bone. The bits of calcium spread out in the vinegar (like sugar dissolves in hot tea). The vinegar doesn't damage the bone's proteins, however. The long, soft proteins left in the bone can be twisted into different shapes or tied in a knot.

One Step Further:

Ask an adult for a hard-boiled egg. Soak the egg in a jar of vinegar for a day or two. What happens to the egg? Why?

Questions:

1. Why do you think some bones take longer to soften in vinegar than others?
2. Some ocean animals have hard shells made of calcium. Can you name three of these animals?
3. Which foods are rich in calcium? Why are these foods good for you to eat?
4. If a baby tooth falls out and you put it in vinegar, what do you think will happen?
5. In the weightless environment of outer space, astronauts don't need as much calcium in their bones as they do on Earth. Why do you think this is?

Colorful Celery Stalks

Can you see water travel through a plant?

Background:

Plants have roots to take up water from the ground. The water is then carried to all parts of the plant through tiny tubes called xylem. Xylem tubes go from the roots through the stem to the leaves and flowers. Almost all plants, from giant oak trees to small buttercups, use xylem tubes to carry water.

What You'll Need:

- measuring cup
- water
- tall glass
- flat surface
- red food coloring
- long spoon
- knife (to be used only by an adult)
- stalk of celery with leaves attached

What to Do:

1. Pour 2 cups water into the glass, then place the glass on a flat surface. Add 15 drops red food coloring. Stir with the spoon.
2. Ask an adult to cut a new, straight edge across the bottom of the celery stalk. Place the celery stalk, cut end down, into the glass of red water.
3. After three hours, take the celery stalk out of the water. Rinse it and look at the cut end. The tips of the xylem tubes will be red.
4. Place the celery stalk back in the water. Check it every few hours for the next couple of days. What happens?

What Happened:

Celery has small xylem tubes. These tubes carry water from the bottom of the plant to the top. When you placed the celery stalk, or stem, into red water, the water was pulled into the stalk's xylem tubes. The red water crept up the xylem tubes in the stalk and entered xylem tubes in the leaves. The red water made the stalk and leaves appear red. The red color helped you see the water travel through the celery plant.

One Step Further:

Get a white carnation with a long stem. Ask an adult to cut the stem in half lengthwise (the flower should not be cut). Pour ¾ cup water into each of two glasses. Mix 6 drops blue food coloring into one glass of water. Mix 6 drops red food coloring into the other glass of water. Place the glasses beside each other on a flat surface. Put one half of the flower stem into the red water and the other half into the blue water. Leave the flower standing in the water for four or five days. What happens?

Questions:

1. What happens to a plant when it doesn't get enough water?
2. Celery seeds are sometimes used to flavor food. Where do you think celery seeds come from?
3. Plants also have tubes called phloem. They carry substances from the leaves to the rest of the plant. What do you think they carry?
4. People have tubes in their bodies also. Can you name some of these tubes? What do they do?
5. Some small, simple plants like algae don't have xylem tubes. How do you think they get water? Do you think a tree could survive without xylem tubes?

Making Magnets

How can you make a magnet?

Background:

Magnets have the power to attract, or pull, certain metal objects toward themselves. This power is called magnetism. Most magnets are made of iron or iron-rich metals, like steel.

What You'll Need:

- three steel paper clips
- flat surface
- bar magnet
- iron nail
- piece of aluminum foil
- quarter
- steel screwdriver
- rock
- scissors
- nickel
- plastic spoon

BE CAREFUL WITH SHARP OBJECTS.

What to Do:

1. Place three steel paper clips on a flat surface. Hold the bar magnet just above the paper clips. The magnet will attract, or pull up, the steel paper clips.
2. Pick up the nail and hold it just above the paper clips. What happens?
3. Now stroke the nail 50 times with one end of the bar magnet. Always stroke in the same direction, and be sure to lift the magnet away from the nail after each stroke.
4. Hold the nail above the paper clips again. Does the nail attract the paper clips? If it does, it has become a magnet.
5. Now repeat steps 2 through 4 with each of the remaining items: a piece of aluminum foil, a quarter, a steel screwdriver, a rock, scissors, a nickel, and a plastic spoon. Which items become magnets?

What Happened:

Magnetism is an invisible force that gives magnets the power to attract certain metal objects, like steel paper clips. Since magnets are made almost entirely of iron, most of their atoms (tiny particles that make up all substances) are iron atoms. Not all iron objects are magnets, however. An iron object is a magnet only if its iron atoms are lined up. At first, the iron nail was not a magnet. It became one when you stroked the nail with the bar magnet, forcing the iron atoms to line up. The same thing happened with the other items made of iron or steel—the screwdriver and the scissors. Objects that are not made of iron contain few or no iron atoms, so they cannot become magnets.

One Step Further:

To make a compass: Rub a bar magnet along a steel sewing needle 50 times. The needle will become a magnet. Tape the needle to a quarter-sized slice of cork. Fill a small plastic bowl with water and place the cork, with the needle attached, into the water. The needle will rotate until it points in a north-south direction. Why does this happen?

Questions:

1. Why can't the aluminum foil, quarter, rock, nickel, and plastic spoon be magnetized?
2. Natural magnets, called lodestones, are made from a mineral called magnetite. What kind of atoms does magnetite contain?
3. What kind of atoms is a copper bowl made of?
4. Some very powerful magnets, called electromagnets, can lift objects that weigh thousands of pounds. What do you think these magnets could be used for?
5. Imagine that you can make a new kind of magnet that attracts things other than iron and steel. What would it attract? What would you use it for? What would you call it?

Electric Balloons

What happens when balloons get extra electrons?

Background:

All substances are made of tiny atoms, or particles. Two kinds of even smaller particles in atoms are protons, which have positive charges, and electrons, which have negative charges. Atoms have equal numbers of protons and electrons, so they are neutral. They have no overall charge. Electrons "jump" on and off atoms quite easily, however. When atoms lose electrons, they become positively charged. When atoms gain electrons, they become negatively charged. Charged atoms are called ions. Things that have similar charges, like two positive ions or two negative ions, repel each other. This means they push each other away. Things that have different charges, like a positive ion and a negative ion, attract each other.

What You'll Need:

- two round balloons
- two pieces of string
- piece of wool clothing, such as a sock, sweater, or scarf

What to Do:

1. Ask an adult to help you blow up each balloon and tie a knot in the end.
2. Tie a piece of string around the end of each balloon.
3. Hold the balloons by the tops of the strings and touch them together. They don't do anything.

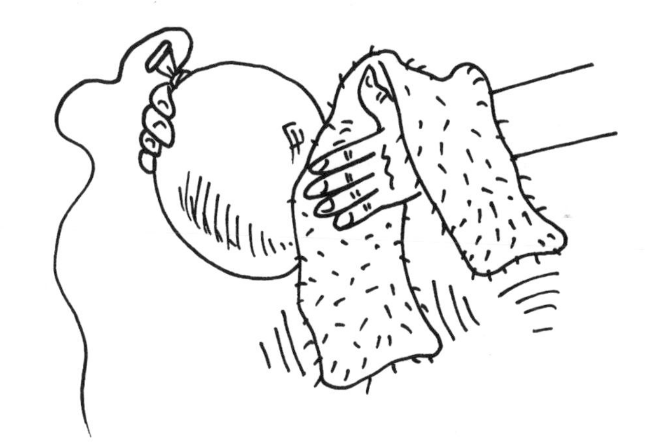

4. Now rub each balloon on the piece of wool clothing.
5. Hold the balloons by the tops of the strings and touch them together again. What do the balloons do this time?

What Happened:

When you first touched the balloons together, their atoms had equal numbers of protons and electrons, so the balloons had no overall charge. They did not attract or repel each other. When you rubbed the balloons on the piece of wool clothing, electrons from the wool cloth "jumped" onto atoms in the balloons. The atoms in the balloons then became negative ions. Since both balloons were negatively charged, they pushed each other away. We call this static electricity.

One Step Further:

Shake some pepper onto a piece of white paper. Get a plastic spoon and rub it with the wool. Hold the spoon over the paper. What happens? Why?

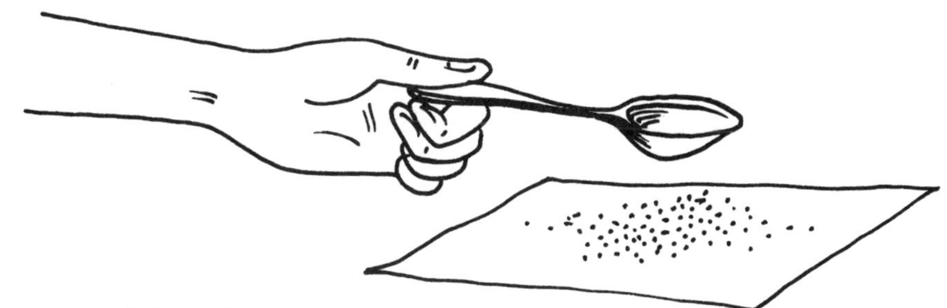

Questions:

1. Which parts of atoms have positive charges?
2. What's the difference between an atom and an ion?
3. After you run a comb through your hair, the comb may attract small pieces of paper. Why do you think this happens?
4. Electric currents, made of moving electrons, run many appliances in your home. Can you name any objects or vehicles capable of carrying people that are powered by electricity?
5. How are helium balloons able to fly?

Water Ring

How can you make a hole in water?

Background:

The molecules, or particles, of a liquid tend to cling together. This is called cohesion. The molecules of a liquid may also cling to other substances. This is called adhesion. Water molecules have strong cohesion. They tend to “stick” together very tightly.

What You’ll Need:

- red food coloring
- ½ cup water
- spoon
- saucer
- flat surface
- eyedropper
- rubbing alcohol

BE CAREFUL WITH RUBBING ALCOHOL.

What to Do:

1. Place 3 drops food coloring into ½ cup water. Mix it with the spoon.
2. Put the saucer on a flat surface. Pour enough colored water into the saucer to just cover the bottom of the saucer.
3. With the eyedropper, place 1 drop rubbing alcohol in the center of the colored water. What happens?

What Happened:

Water molecules cling to each other very strongly. Because of this, a “skin” created by surface tension forms on the surface of the water (see “Scatter the Pepper”). Water molecules do not adhere, or cling, to alcohol molecules. When you dropped alcohol into the water, the “skin” of the water pulled away from the alcohol. Since the alcohol was in the center of the saucer, the water formed a ring around the alcohol.

One Step Further:

Poke two small holes beside each other in the bottom of a paper cup. Hold the cup over the sink, and fill it with water. Two small streams of water will pour out, one from each hole. Pinch the two streams together with your fingers. Do the two streams stick together or come apart? Why does this happen?

Questions:

1. Do you think the molecules in a solid show cohesion?
2. Do you think the molecules in a gas, like air, show cohesion?
3. When you wet two pieces of paper, they will stick together. When the papers dry, they will come apart. Why does this happen?
4. When it rains, water doesn’t just drain away. Some of it adheres to bits of soil. What does “adhere” mean? How does this water help plants?
5. Cut two or three colorful magazine pages into interesting shapes. Make a picture by pasting, or “adhering,” the shapes onto a large sheet of paper. Why do we say paste is an adhesive?

Scatter the Pepper

How can dish detergent move pepper across a bowl of water?

Background:

A container of water contains trillions of water molecules, or particles. The water molecules are linked together (see "Water Ring"). They pull on each other in all directions. At the surface of the water, however, the linked molecules are only pulled to the sides and down. They are not pulled up because there are no water molecules above them. The surface molecules, therefore, form a strong, stretchy "skin" across the top of the water. This is called surface tension.

What You'll Need:

- bowl
- water
- flat surface
- black pepper in a shaker
- dish detergent
- toothpick

What to Do:

1. Fill the bowl almost to the top with water and place it on a flat surface.
2. Sprinkle a few shakes of pepper into the bowl of water. The pepper will spread out over the water's surface.

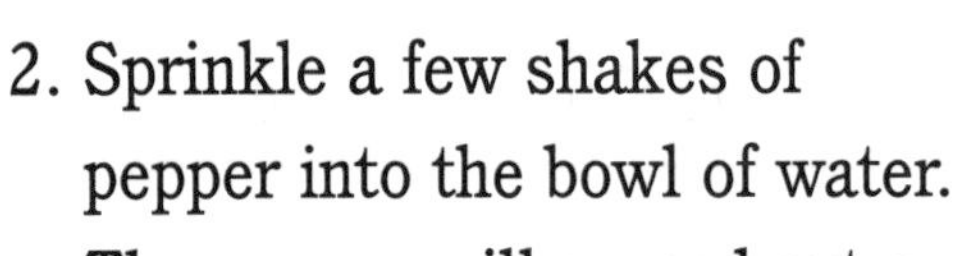

3. Place a drop of dish detergent onto the end of the toothpick.
4. Hold the soapy end of the toothpick over the center of the bowl and lightly touch the surface of the water. What happens?

Note: You must wash all the soap out of the bowl and use fresh water if you want to repeat the experiment.

What Happened:

The pepper floated on the water because of surface tension. Surface tension is created by linked water molecules pulling on each other. Soap breaks the links between water molecules. When you dipped dish detergent, a type of soap, into the center of the bowl, the detergent broke the links between the water molecules in the bowl's center. The water molecules around the rim of the bowl, however, remained linked and continued to pull. They pulled the water, and the floating pepper, to the sides of the bowl. That is why the pepper "shot" to the edges of the bowl.

One Step Further:

Sprinkle a little water on a piece of waxed paper. The water will form round droplets. Use a toothpick to move the droplets around on the paper. Do they remain round or lose their shape? Why?

Questions:

1. What would happen if you placed a penny on the surface of water? Why?
2. Soap helps wash away dirt. Can you name some types of soap, other than dish detergent?
3. What do we mean when we say things are linked together? Can you name two things that are linked together?
4. How is the "skin" on the surface of water like your skin? How is it different from your skin?
5. Water bugs are a group of insects. Some are able to run on top of water. How are they able to do that?

Fungus Fun

Can you grow a fungus in your home?

Background:

Molds are a type of fungus. Molds usually grow on things that come from dead plants and animals, like rotten logs, paper, leather, and old food. All fungi absorb, or take in, food from the things they grow on. Fungi produce tiny spores that grow into new fungi. Spores are easily spread by wind and water, and are found everywhere. They're in buildings, houses, forests, fields, ponds, rivers, and so on.

What You'll Need:

- saucer
- flat surface
- 2 teaspoons water
- slice of bread
- zip-top sandwich bag
- warm, dark place

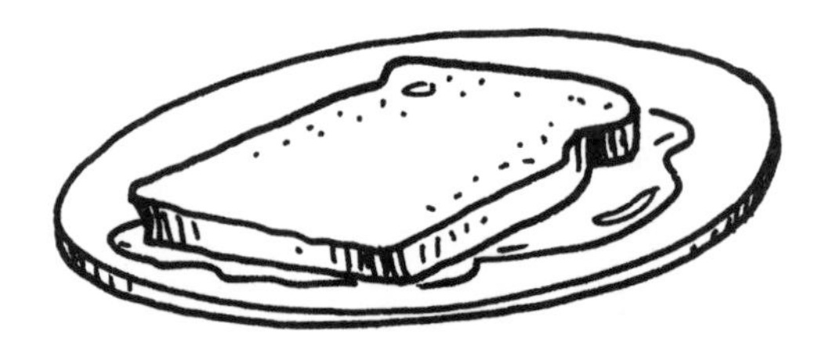

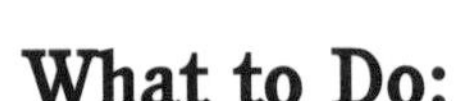

What to Do:

1. Put the saucer on a flat surface. Pour 2 teaspoons water into the saucer and spread the water around.
2. Lay the bread in the wet saucer. Make sure the bread gets very moist.
3. Remove the bread from the plate and put it in the sandwich bag. Seal the bag tightly.
4. Put the bag in a warm, dark place, like a kitchen cupboard. After one or two weeks, check the bread. There should be a layer of mold growing on it.

What Happened:

Although you can't see them, mold spores are everywhere, including in your home. When you prepared the slice of bread for the experiment, mold spores from the air got onto the bread. When you wet the bread and put it in a cupboard, you created moist, warm conditions that helped the spores sprout. After a week or two, the spores grew into the fuzzy mold that covered the bread.

One Step Further:

Mushrooms are a type of fungus. Ask an adult to help you look for mushrooms, molds, and other fungi growing in a yard, in a field, or along a nature trail. Take along paper and crayons and draw pictures of the fungi you see. NEVER EAT WILD MUSHROOMS—THEY MAY BE POISONOUS.

Questions:

1. What did the mold you grew use for food?
2. If you had put the moist bread in the refrigerator, do you think the mold would have grown faster or slower? Why?
3. Why shouldn't you eat mushrooms that you find outside?
4. Fungi are *not* plants. How do you think they're different from plants?
5. People sometimes eat fungi. For example, blue cheese contains molds, bread is made with yeast (which is a fungus), and many recipes call for mushrooms. On a separate piece of paper, write a menu for a meal that contains several kinds of fungi. Draw a picture of the meal. Would you eat this meal?

Goopy Goo

What is a colloid and how does it behave?

Background:

A wooden block is a solid. It doesn't change its shape. Water is a liquid. It takes the shape of any container that holds it. Some substances can act like *both* solids and liquids. These substances are called colloids.

What You'll Need:

- small bowl
- flat surface
- measuring cup
- cornstarch
- water
- spoon

What to Do:

1. Put the bowl on a flat surface.
2. Put 1 cup cornstarch into the bowl.
3. Add ½ cup water, then stir.
4. The mixture should be fluid but hard to mix. If the mixture is too thick, add a little water. If the mixture is too watery, add a little cornstarch. This substance is a colloid. It's your "Goopy Goo."
5. Scoop out a handful of Goopy Goo and roll it between your palms. It will become stiff, like a solid.
6. Hold the stiff blob of Goopy Goo in your hand. Put your hand over the bowl and spread your fingers. The Goopy Goo will drip into the bowl, like a liquid.

What Happened:

Goopy Goo is a colloid. It is made of solid particles of cornstarch suspended, or "floating," in water. When you squeezed the Goopy Goo, the cornstarch particles were pushed together. The colloid then behaved like a solid. When you allowed the Goopy Goo to rest lightly in your hand, the cornstarch particles spread out. The colloid then behaved like a liquid.

One Step Further:

Try the following experiments with the Goopy Goo and see what happens.

- Cut the Goopy Goo in the bowl with a plastic knife. Does it split into separate chunks?
- Let your fingers sink into the bowl of Goopy Goo. Is it hard to pull your fingers back out?
- Try to stir the Goopy Goo quickly with your finger. What happens? Now try to stir the Goopy Goo slowly with your finger. What happens now?
- Slap the Goopy Goo hard with a wooden spoon. Does it splash?
- Slowly try to pour the Goopy Goo into a plastic cup. Does it flow?
- Put a few drops of food coloring into your Goopy Goo and stir it in slowly. What happens to the color of the Goopy Goo?
- Put Goopy Goo into a zip-top sandwich bag and seal the bag shut. Squeeze the Goopy Goo. Does it change shape?

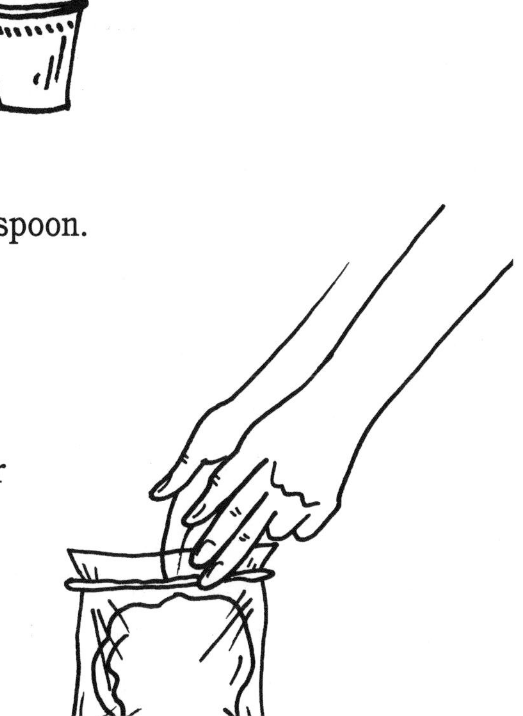

Note: When you are finished with your experiments, let the Goopy Goo become firm and throw it in the trash. Don't wash it down the sink. It can clog the drain.

Questions:

1. Liquids come in containers of many shapes and sizes. Can you describe some of these containers? What do they hold?
2. If you slapped a liquid, like water, with a wooden spoon, what do you think would happen?
3. How can a solid, like a rock, become a liquid?
4. Do you think orange juice is a colloid? Why or why not?
5. Quicksand is a colloid that behaves like Goopy Goo. In movies, people sometimes sink into pools of quicksand. They usually struggle to get out and eventually drown. If a person really fell into a pool of quicksand, what do you think would be the best way to escape?

Answers

Pages 8–9

OSF: The force of gravity on the pencil and the paper is equal. However, on Earth, objects fall through air. Air molecules, or particles, bump into falling objects and slow them down. The air slows down the light, flat piece of paper more than it does the pencil. That is why the paper falls more slowly than the pencil. When the paper is crumpled into a ball, the air doesn't slow it down as much. That is why the crumpled paper and the pencil fall at the same speed. If the pencil and the open paper were dropped in a vacuum, an area with no atmosphere, they would fall at the same speed.

1. Objects fall down because they're pulled to Earth by gravity.
2. You would need a very fast spaceship to overcome Earth's gravity.
3. Answers will vary. All the other planets in our solar system have stronger gravity than Pluto. The other planets are Mercury, Venus, Earth, Mars, Jupiter, Saturn, Uranus, and Neptune.
4. You could jump higher on the Moon. The Moon is smaller than Earth, so gravity is weaker there. The Moon wouldn't "pull" on you as hard as Earth does, so you could jump higher.
5. Answers will vary. Scientists guess that objects pulled into a black hole are crushed into extremely tiny, very heavy specks.

Pages 10–11

OSF: A quarter has more mass than a penny. Therefore, the quarter transferred enough momentum to move two pennies off the row.

1. No, a book lying on your desk doesn't have momentum. Only moving objects have momentum.
2. An elephant has greater mass than a mouse.
3. If a car speeds up, its momentum increases.
4. A big train traveling at 50 miles per hour has greater momentum than a small car traveling at 50 miles per hour because momentum depends on mass and speed.
5. Answers will vary. Sample answer: If one person bumps into another, the second person gets pushed along.

Pages 12–13

OSF: The cup and the marble were both moving across the table. When you stopped the cup suddenly, the marble kept going. The marble, which was not acted upon by a force, continued to move because of inertia.

1. If the coin had been glued to the card, it would have flown off with the card.
2. Your belongings slide off the seat because of inertia. They are moving objects that keep moving, even though the car has stopped.
3. The card didn't keep on going because forces stopped it. It collided with, or bumped into, air molecules, which slowed it down. The card was also pulled down by gravity and finally stopped when it hit the table.
4. Answers will vary. Sample answers: the Sun, Moon, Earth (and other planets), stars, and galaxies.
5. Answers will vary. Sample answers: Motionless objects might start to roll around or fly into the air. Moving objects might stop suddenly or turn around and go the other way.

Pages 14–15

OSF: The ball is round, so it rolls easily. The egg is an oval shape, so it doesn't roll easily. Oval objects don't roll well because they are lopsided. An egg's oval shape might help to keep it from rolling out of a bird's nest.

1. No, the ball will not start spinning again, because it is solid.
2. A compass needle points north.
3. The top collides with air molecules, which stop the top from spinning.
4. Answers will vary. Sample answers: the Sun, the planets, or the Moon.
5. ***Parent:*** Check child's answer.

Pages 16–17

OSF: When you hum into the open end of the tube, the waxed paper at the other end vibrates. This produces a funny buzzing sound.

1. Most sounds you hear come through air.
2. Answers will vary. Sample answers: Sounds in the home include those from a television, a telephone, a radio, people talking, a refrigerator motor, and a washing machine. Sounds in the street include those from a car horn, an ambulance siren, a dog barking, thunder, and a crossing-guard whistle.
3. Sound waves are stronger in water because water molecules are closer together than air molecules. Molecules in liquid are closer together than they are in gases, but they are still farther apart than they are in solids.
4. Answers will vary. Sample answers: They could use radio transmitters (radio waves can travel through space). They could use sign language.
5. Answers will vary. Sample answers: Several different languages may have developed from one language that was spoken a long time ago. Languages may "borrow" words from each other.

Pages 18–19

OSF: When your finger covers the straw, the air pressure pushing up on the water in the straw is greater than the air pressure pushing down on the water in the straw. This prevents the water from flowing out. When you lift your finger, the air pressure pushing down becomes the same as the air pressure pushing up. The water then flows out because of the force of gravity, which pulls everything to the Earth's surface.

1. On a windy day, you can't really see air. However, you can see air blowing things around, like leaves, papers, hats, and clouds.
2. Air pressure doesn't prevent you from walking forward because the air behind you is pushing just as hard as the air in front of you.
3. When suction cups are pressed against a flat surface, the air pressure on top of the cups is greater than the air pressure underneath them. Because of that, the cups stay in place.
4. Water would have more pressure because water is heavier than air.
5. As altitude increases, air pressure decreases. The pilot could judge how high the plane is by looking at the amount of air pressure shown on the barometer.

Pages 20–21

OSF: As you raise the straw, the sound gets lower in pitch. This happens because the column of air in the straw gets longer, and the sound waves get longer (see "Spoon Sounds"). Longer sound waves produce sounds with lower tones. As you lower the straw, the sound gets higher in pitch. This happens because the column of air in the straw gets shorter, and the sound waves get shorter. Shorter sound waves produce sounds with higher tones. This is how a musical instrument called a slide trombone works.

1. When something is compressed, it is squeezed into a small space.
2. Answers will vary. Sample answers: cotton ball, foam-rubber ball, sponge, pillow, or slice of bread.
3. It would have been easier to poke a hole in a baked potato because a baked potato is softer than a raw potato.
4. It's easier to compress a gas than a liquid or a solid because the molecules in a gas are very spread out. There's a lot of space between the gas molecules. The molecules can squeeze into those spaces if they need to.
5. Answers will vary. Sample answers: Machines that use gasoline-powered motors include cars, boats, buses, motorcycles, trucks, and airplanes. Machines with motors powered by electricity include washing machines, clothes dryers, vacuum cleaners, refrigerators, dishwashers, and blenders.

Pages 22–23

OSF: The hot water heats the bottle. After the water is poured out, the warm bottle heats the air inside. When the warm bottle, with the balloon attached, is put into cold water, the air in the bottle cools and contracts. As the air in the bottle contracts, the outside air pushes in. The outside air thrusts the balloon into the bottle and inflates it.

1. The balloon would shrink because the air inside it would contract.
2. If the quarter didn't completely cover the opening of the bottle, the air could slowly escape without making the quarter jump.
3. Answers will vary. Sample answers: rubber band, balloon, accordion, ponytail holder, elastic waistband, and lungs.
4. All substances expand when they are heated. Hot running water heats the jar's lid. The lid expands and loosens up a little, making it easier to unscrew.
5. Answers will vary.

Pages 24–25

OSF: As the bag cools, the water vapor molecules slow down to form water. As the water continues to cool, the molecules slow down even more, to form ice.

1. Yes, your body is made of matter.
2. The chocolate bar melts because its molecules get heated up. This makes the molecules move faster and farther apart, until the chocolate becomes liquid.
3. Answers will vary. Sample answers: solids—ball, book, hot dog; liquids—soda, juice, milk; gases—air, carbon dioxide, helium.
4. The red-hot liquid is called lava. It cools to form rocks.
5. Answers will vary.

Pages 26–27

1. A person's body can usually float because it's less dense than water.
2. The ship's weight is spread over a very large volume. Therefore, it is less dense than water and floats. The marble's weight is contained in a very small volume. Therefore, it is more dense than water and sinks.
3. The extra salt in the Dead Sea increases its weight. A gallon of Dead Sea water weighs more than a gallon of ocean water. Therefore, the Dead Sea is more dense than the ocean.
4. When an iceberg tore a hole in the side of the *Titanic*, water entered the ship. This increased the ship's weight and density. Eventually, the *Titanic*'s density became so great that it sank.
5. A person floats in water because the human body is less dense than water. A person floats in space because there is no gravity holding the person down.

Pages 28–29

OSF: The honey, water, and oil don't mix. They form three layers, arranged in order of density. The oil is the least dense, so it floats on top. The honey is the most dense, so it will be on the bottom. Water's density is between that of honey and oil, so it stays in the middle.

1. Food coloring molecules are not very different from water molecules, so they can mix together.
2. The oil floats on the surface of the ocean. It is very hard to clean up.
3. Answers will vary. Sample answers: Oil is used for frying food, for baking, to stop door hinges from squeaking, to make machines run smoothly, to make gasoline, and to make plastics.
4. It means they are very different from each other and don't get along.
5. You couldn't clean your hands very well by running plain water over them. Soap helps water stick to oil so that it can be washed off your skin.

Pages 30–31

OSF: When you looked into the bowl of the spoon, your image was upside down and a little distorted. When you looked into the back of the spoon, your image was right side up and a little distorted. This happened because curved surfaces don't reflect light rays straight back to your eye, like a regular mirror does. Surfaces that curve in or out bend light in different ways. This makes images look distorted.

1. Yes, light bends whenever its speed changes.
2. Answers will vary. Sample answers: a straw, your arm, your leg, a licorice stick, a pipe cleaner, a leaf, and a twig.
3. Yes, it would look bent.
4. You can't see in the dark because there's no light to bounce off objects and enter your eyes.
5. To figure out how far light travels in a minute, multiply 186,000 by 60 (because there are 60 seconds in a minute). To figure out how far light travels in an hour, multiply the previous answer by 60 (because there are 60 minutes in an hour).

Pages 32–33

OSF: When the circle spins quickly, you can't see each color separately. Your brain blends the colors into a whitish blur, which is the result of all the colors mixing together. This is similar to what happens in white light.

1. You are most likely to see a rainbow after it rains because water drops remain in the atmosphere. The water drops refract sunlight.
2. Orange light has longer waves than green light.
3. Answers will vary. Sample answers: heat, solar energy, geothermal energy, atomic energy, and electricity.
4. Answers will vary. Sample answers: Sir Isaac Newton discovered the law of gravity and the laws of motion.
5. Answers will vary. Sample answers: a brown moth on a tree trunk, a green lizard on a leaf, and a tiger in tall grass.

Pages 34–35

OSF: Because of the way your eyes work, you sometimes see things that aren't really there. When you stared at the pink circle, your eyes got tired of seeing pink. When you shifted your eyes to the blank paper, your eyes saw a green circle. Green is the complementary color of pink.

1. The sky is black at night because that side of the Earth is facing away from the Sun, and therefore receives less light.
2. Answers will vary. Sample answers: Green light bounces off a dollar bill, purple light bounces off a plum, and blue light bounces off a blue backpack.
3. Water gets into the atmosphere by evaporating from oceans, lakes, and other bodies of water. Dust is carried into the atmosphere by wind.
4. Answers will vary. Sample answers: earthquake (seismic) waves, water waves, sound waves, radio waves, brain waves, and microwaves.
5. Answers will vary.

Pages 36–37

1. The primary colors are red, yellow, and blue. The secondary colors are orange, purple, and green.
2. To make lavender, you would mix red, blue, and white paint, or purple and white paint.
3. Answers will vary. Sample answers: Tomatoes change from green to red; bananas change from green to yellow; peppers change from green to red (if left on the vine); pumpkins change from green to orange.
4. A person who is color blind cannot see colors properly. This is usually caused by a defect in the retina or nerves in the eye.
5. Answers will vary.

Pages 38–39

1. Answers will vary depending on the brand of marker.
2. Beets—red; grapes—purple; spinach—green; onion skins—yellow or brown; blueberries—purple or blue.
3. When you place a sponge in water, the water creeps into the sponge. This is the same process explained in the experiment you just did. It is called capillary action.
4. The 96 colors are created by mixing together the three basic colors—red, yellow, and blue—in different amounts (see "Color Combos"). White is used to make lighter shades.
5. Answers will vary.

Pages 40–41

OSF: Your body normally has a temperature of 98.6°F. The hand with the glove stayed warm longer because your body warmth was trapped inside the glove. This kept your hand from feeling cold for a little while. The hand without the glove became cold sooner because your body warmth quickly escaped into the cold water.

1. Answers will vary. Sample answers: Foods that taste best hot are hamburgers, french fries, pizza, corn on the cob, and chicken soup. Foods that taste best cold are ice cream, juice, yogurt, milk, and green salad.
2. The doctor uses a thermometer. He or she is trying to find out if your temperature is higher than normal.
3. Answers will vary. Sample answers: Hot cocoa has steam rising from it, a hot person perspires, and a barbecue grill contains glowing charcoal; cold grass has a coating of frost, a cold person shivers, and a cold roof has icicles hanging from it.
4. Your other senses are sight, hearing, smell, and taste.
5. On a hot day, the thermometer would show a higher temperature; the red line would be higher. On a cold day, the thermometer would show a lower temperature; the red line would be lower.

Pages 42–43

1. A carrot reflects orange light. A beet reflects red light. A white pearl reflects all the colors of light. A black marble absorbs all the colors of light and does not reflect any colors.
2. Black clothes absorb all the colors of light. The light's energy warms up the clothes—and you!
3. Your body normally has a temperature of 98.6°F, and it gives off heat. Mittens trap the heat and keep it close to your hands. This helps keep your hands warm.
4. Sunlight is so bright that it can damage your eyes if you look directly at the Sun.

5. Answers will vary. Sample answers: Places close to the equator, like Mexico, Brazil, Indonesia, and Kenya, have a hot climate. Places near the Poles, like Alaska, Greenland, the Arctic, and Antarctica, have a cold climate.

Pages 44–45

OSF: The eggshell is made of hard calcium minerals. The vinegar dissolves, or breaks up, the calcium minerals. When all the calcium minerals are dissolved, the eggshell "disappears."

1. The amount of time it takes a bone to soften depends on the size of the bone. Bigger bones have more calcium, so they take longer to soften.
2. Answers will vary. Sample answers: clams, oysters, mussels, snails, and conchs.
3. Answers will vary. Sample answers: Dairy products, such as milk, and leafy green vegetables are rich in calcium. They're good for you because you need calcium for strong bones and teeth.
4. Teeth contain calcium. Vinegar would take the calcium out of the tooth, and the tooth would get soft.
5. In the weightlessness of outer space, astronauts float around. They don't need to stand on their feet or walk, so their bones don't need to support much weight. In Earth's gravity, astronauts' bones must be strong enough to support their weight when they stand and walk.

Pages 46–47

OSF: The colored water was sucked into the xylem tubes in the carnation stem. The red water traveled up one side of the stem to one side of the flower, making the flower petals on that side red. The blue water traveled up the other side of the stem to the flower, making the flower petals on that side blue.

1. Without water, a plant wilts, or droops.
2. Celery seeds are made by celery plants. The seeds are found in the fruits of the plants. You usually don't see the fruits, because celery sold in grocery stores contains only stems and leaves.
3. Phloem tubes carry food, most of which is made in the plant's leaves, to the rest of the plant.
4. Answers will vary. Sample answers: People have arteries and veins that move blood. People have a digestive tube that breaks up food.
5. Answers will vary. Sample answers: Algae live in water. Their cells (the tiny parts that make them up) absorb water directly from their habitat. A tree could not survive without xylem tubes. Trees are too large to take in water the way algae and some other simple plants do.

Pages 48–49

OSF: The needle points in a north-south direction because one pole of the needle (called the "north-seeking pole") points to the Earth's North Pole. The other pole of the needle points in the opposite direction, to the Earth's South Pole.

1. The aluminum foil, quarter, rock, nickel, and plastic spoon can't be magnetized because they're not made of iron.
2. Magnetite contains iron atoms.
3. A copper bowl is made of copper atoms.
4. Answers will vary. Sample answers: picking up broken cars and moving steel beams used for building skyscrapers.
5. Answers will vary.

Pages 50–51

OSF: When you rubbed the plastic spoon with the piece of wool clothing, the spoon became negatively charged. It then attracted the positively charged pepper. That is why the pepper jumped up to the spoon and stuck to it.

1. Protons have positive charges.
2. An atom has no overall charge. An ion is an atom that has a positive or negative charge.
3. Electrons from your hair "jump" onto the comb. The comb becomes negatively charged. It then attracts positively charged bits of paper.
4. Answers will vary. Sample answers: trains, electric cars, elevators, escalators, and trolleys.
5. Helium balloons fly because helium gas, which is lighter than air, rises.

Pages 52–53

OSF: The cohesion of the water molecules makes the two streams of water stick together.

1. The molecules of a solid do show cohesion. They stick together.
2. The molecules of gas don't show cohesion. They spread apart.
3. When you wet the papers, there is water in each paper. The water sticks together due to cohesion, so the papers stick together. When the water evaporates, the papers become dry. There's no more water to stick together, so the papers come apart.
4. *Adhere* means "stick together." Water adheres, or "sticks," to bits of soil. This water can then be taken up by plants, which need water to live.
5. Paste, or glue, is an adhesive because it makes things stick together.

Pages 54–55

OSF: In the center of each droplet of water, the water molecules pull on each other in all directions. At the surface, however, the water molecules are only pulled into the droplet (toward the other water molecules). This creates very strong surface tension that keeps each droplet of water round.

1. The penny would sink. It is heavy enough to break the surface tension of the water.
2. Answers will vary. Sample answers: laundry soap, bath soap, shampoo, car wash soap, and bathroom cleanser.
3. Things that are linked together are connected. Rest of answers will vary. Sample answers: links of a chain-link bracelet, cars of a train, cities connected by a bridge, and people talking by telephone.
4. Answers will vary. Sample answers: The water's "skin" is like your skin because it's on the surface of water and your skin is on the surface of your body. The water's "skin" is unlike your skin because it can be broken easily and your skin cannot be broken easily.
5. Surface tension allows some water bugs to run on top of water.

Pages 56–57

1. The mold used the bread for food.
2. The mold would have grown slower in the refrigerator because molds don't grow as well in cold places.
3. Some mushrooms are poisonous and may harm you. It would be difficult to tell which ones are safe to eat.
4. Answers will vary. For example: Fungi absorb food (they don't make their own food like plants do); fungi aren't green, don't have flowers and leaves, and don't grow from seeds.
5. Answers will vary.

Pages 58–60

OSF: When you cut the Goopy Goo, it splits into separate chunks, then immediately flows back together. It is hard to pull your fingers out of the Goopy Goo, especially if you try to do it quickly. It is hard to stir the Goopy Goo if you move your finger quickly, but you can do it if you move your finger slowly. The Goopy Goo does not splash if you slap it with a wooden spoon. The Goopy Goo will flow slowly into a cup if you try to pour it. The Goopy Goo will change color if you slowly stir in food coloring, although the color may look light, or subdued. The Goopy Goo can change shape when you squeeze it in the zip-top sandwich bag, but it will immediately spread out again.

1. Answers will vary. Sample answers: Car oil comes in cylindrical cans or plastic bottles; milk comes in gallon jugs or quart containers; fruit juice comes in bottles or cans.
2. If you slapped water with a wooden spoon, it would probably splash because water is a liquid.
3. When a rock is heated to very high temperatures, it can become a liquid.
4. Orange juice is not a colloid because it can not become solid. It is always a liquid.
5. The best way to escape from a pool of quicksand is by camly, slowly floating on your back with your arms outstretched at right angles to your body. This way, you can slowly roll off the quicksand onto solid ground. The slower your movements, the less "solid" the quicksand will be.

Index

A
adhesion, 52–53, 63
air molecules, 16–17, 18
air pressure, 18–19, 23, 61
algae, 63
antimatter, 25
atoms, 49, 50, 51

B
barometer, 19
blue light, 35
body temperature, 62

C
calcium, 44, 45, 63
capillary action, 39
cohesion, 52–53, 63
colloids, 58–59, 63
color, 36–37, 38–39, 42, 62
color blindness, 62
compass, 49, 63
compressed air, 20–21
contraction, 22–23, 61

D
density, 26–27, 28–29, 30, 62
distortion, 62

E
Earth, 9, 18
electric current, 51
electrons, 50, 51
energy, 8, 10–11, 33, 42–43
expansion, 22–23, 61
experiment guidelines, 7

F
fungi, 56–57, 63

G
gases, 17, 24–25
gravity, 6, 8–9, 13, 61

H
habitat, 33
helium gas, 63
hypothesis, 5

I
inertia, 12–13, 14–15, 61
ions, 50, 63
iron atoms, 49

L
laws of nature, 6
light, speed of, 30–31, 62
light energy, 42–43
light waves, 32–33, 34–35, 62
liquids, 24–25, 28, 52, 58, 59
long waves, 35

M
magnetism, 48–49, 63
mass, 10–11, 61
matter, forms of, 24–25
measurement, 6
metric measurement, 6
minerals, 44–45, 63
molds, 56–57, 63
molecules, 10, 16–17, 18, 29, 52–53, 54–55, 61
momentum, 10–11, 61
mushrooms, 57, 63

N
negative charge, 50, 63
neutral charge, 50
Newton, Isaac, 6, 33, 62

P
phloem tubes, 47, 63
pitch, sound, 61
plants, 46–47
positive charge, 50, 63
primary color, 36, 37, 62
proteins, 44–45
protons, 50, 51

Q
quicksand, 60, 63

R
rainbow, 32, 33
red light, 33, 35
reflection, 62
refraction, 30–31, 33, 62
roots, plant, 46–47

S
scientific method, 5
scientific principle, 6
secondary color, 35, 36, 62
short waves, 35
solids, 17, 24–25, 58, 59
sound waves, 16–17, 61
speed, 10–11, 61
spinning, 14–15
spores, mold, 56–57
static electricity, 50–51
sunlight, 32–33, 34–35
surface tension, 53, 54–55, 63

T
temperature, 40–41, 62
thermometer, 40, 62

V
violet light, 33
volume, 28, 61

W
water molecules, 29, 52–53, 54–55, 63
water vapor, 25
white light, 32–33, 62

X
xylem tubes, 46–47, 63

Y
yeast, 57